DEDICATION

This book is dedicated to our loving daughter Jayani.

ACKNOWLEDGEMENTS

We express our gratitude to our parents and in-laws for their constant encouragement, support and blessings.

It will be an injustice if we do not thank all our students for their innovative ideas and feedback.

CONTENTS

LEGAL PROCEDURE

1. Oath is not administered to
    A. The children below the age of 12 years
    B. The doctors as they are expert witnesses
    C. An atheist as he does not believe in the existence of God
    D. Illiterates as they do not understand the meaning of oath

2. Evidence that does not proceed from the personal knowledge but from the mere repetition of what one has heard is known as
    A. Circumstantial evidence
    B. Documentary evidence
    C. Hearsay evidence
    D. Oral evidence

3. The term 'Plaintiff' connotes
    A. Assailant
    B. Complainant
    C. Defendant
    D. Cross examiner

4. Leading questions are permitted during
    A. Examination-in-chief
    B. Cross-examination
    C. Re-examination
    D. All of the above

5. A doctor willfully telling a lie under oath in the court of law is called
    A. Perjury
    B. Professional misconduct
    C. Professional negligence
    D. Re judicata

[**Answers:** 1 - A, 2 - C, 3 – B, 4 - B, 5 - A]

6. Not obeying the summons in civil case renders the witness to an action of
    A. Fine
    B. Imprisonment
    C. Paying damages
    D. All of the above

7. Which one of the following type of evidence is most valid?
    A. Circumstantial
    B. Documentary
    C. Oral
    D. Hearsay

8. Panchanama means
    A. Five witnesses
    B. Five guiding principles
    C. Five aspects of a murder
    D. The inquest report

9. Conduct money is the money
    A. Given to conduct the inquiry
    B. Given to the surgeon to conduct the surgery
    C. Issued to attend to the civil court
    D. Issued to attend the summons from the criminal court

10. The offence of perjury is committed when a person
    A. Does not answer the defense council
    B. Hurls a shoe at the sitting judge
    C. Ignores the summons of sessions court
    D. Willfully utters false statement under oath

[**Answers:** 6 - C, 7 - C, 8 - D, 9 - C, 10 - D]

11. The dying declaration must be sent to the
    A. Medical superintendent
    B. Magistrate
    C. Police superintendent
    D. Public prosecutor

12. If the person survives after giving dying declaration, then
    A. The same is accepted as evidence
    B. The person has to appear before the court and sign the declaration
    C. The person has to appear before the court and give oral evidence
    D. The declaration is to be discarded

13. Which one of the following statement regarding is cross-examination not correct?
    A. Is done after the examination-in-chief
    B. Is done by the defense lawyer
    C. Permits questions that have suggested answers
    D. Should be completed within a specified time

14. Which one of the following magistrate holds the inquest?
    A. First class magistrate
    B. Sub divisional magistrate
    C. District magistrate
    D. Executive magistrate

15. In which one of the following cases police does not conduct the inquest?
    A. Accidental deaths
    B. Dowry related deaths
    C. Homicidal deaths
    D. Suicidal deaths

[**Answers:** 11 – B, 12 – C, 13 – D, 14 – D, 15 – B]

16. In which of the following magistrate's inquest is mandatory?
    A. Custodial deaths
    B. Exhumation
    C. Dowry death
    D. All of the above

17. Police officer is empowered to arrest a person without a warrant in
    A. Cognizable offence
    B. Warrant case
    C. Summon case
    D. Non-cognizable offence

18. Which one of the following is not an example of cognizable offence?
    A. Dowry death
    B. Forgery
    C. Murder
    D. Rape

19. The lawyer who argues in the court of law on behalf of the police is the
    A. District lawyer
    B. Defence lawyer
    C. Public prosecutor
    D. Police lawyer

20. Evidentiary materials that are produced in the court during the trial are known as
    A. Corpus delicti
    B. Evidentiary objects
    C. Exhibits
    D. Crime findings

[**Answers:** 16 – D, 17 – A, 18 – B, 19 – C, 20 – C]

21. Who amongst the following has the privilege of professional secrecy?
    A. Doctors
    B. Lawyers
    C. Police
    D. Health minister

22. Which one of the following statements in relation to civil cases is not correct?
    A. Punishment is as per the Indian Penal Code (IPC)
    B. The dispute is between two private parties
    C. The victim is represented by the public prosecutor
    D. Usually guided by the law of torts

23. The type of evidence that carries that amount of proof sufficient to establish the fact in question is called
    A. Conclusive evidence
    B. Direct evidence
    C. Hearsay evidence
    D. Prima facie evidence

24. In which one of the following criminal cases, the doctor is not legally bound to inform the police?
    A. Attempted suicide
    B. Homicidal poisoning
    C. Road traffic accident
    D. Sexual assault

25. Coroner's court was present in
    A. Mumbai
    B. Delhi
    C. Chennai
    D. Bangalore

[**Answers:** 21 – B, 22 – C, 23 – D, 24 – A, 25 - A]

MEDICAL ETHICS AND LAW

1. Which of the following is not the function of the Indian Medical Council?
    A. Issue the warning notice
    B. Maintain the medical register
    C. Prescribe standard for the post graduate education
    D. Take legal action for infamous conduct

2. For which one of the following act warning notice may be issued?
    A. Adultery
    B. Advertisement
    C. Dichotomy
    D. None of the above

3. Which one of the following is not an ingredient of Professional Negligence'?
    A. Duty
    B. Dereliction
    C. Damages
    D. Direct causal connection

4. Which one of the following is an example of 'Res Ipsa Loquitur'?
    A. Burns caused by fomentation
    B. Forceps in the abdomen after surgery
    C. Transfusion of wrong blood group
    D. All of the above

5. All of the following statements are true in a case of professional misconduct except
    A. Disciplinary action taken by the State Medical Council
    B. Damage to the patient must be proved
    C. Violation of the code of medical ethics
    D. Punishment include professional death sentence

[Answers: 1 - D, 2 - D, 3 - C, 4 - D, 5 - B]

6. The latin phrase for 'The thing speaks for itself' is
    A. Actus Novus Intervenience
    B. Res Judicata
    C. Res Ipsa loquitur
    D. None of the above

7. Which one of the following is both illegal and infamous conduct?
    A. Dichotomy
    B. Advertisement
    C. Using 'Red Cross' emblem by doctors on their vehicles
    D. Covering

8. A doctor while exercising discretion can withhold some information from the patient under the doctrine of
    A. Professional secrecy
    B. Res ipsa loquitur
    C. The law of full disclosure
    D. Therapeutic privilege

9. Disciplinary action by the governing council can be taken if the doctor is found guilty of
    A. Contributory negligence
    B. Criminal negligence
    C. Violating the code of ethics
    D. Wrong diagnosis

10. Medical confidentiality can be broken under the following circumstances except
    A. To discuss with another doctor for the benefit of the patient
    B. As a part of the statutory requirement
    C. When the patient does not pay the doctor's fees
    D. In most of the countries with the patient's permission

[**Answers:** 6 - C, 7 - C, 8 - D, 9 - C, 10 - C]

11. Which one of the following is an example of professional misconduct?
    A. Attending to the patient under the influence of alcohol
    B. Leaving the instrument in the body cavity after the surgery
    C. Operating on the wrong limb of the patient
    D. Performing a certain surgery to which the patient has not consented for

12. Under which one of the following doctrines, the doctor who may not actually be directly negligent will be held responsible for negligence?
    A. Contributory negligence
    B. Res Ipsa Loquitur
    C. Vicarious liability
    D. None of the above

13. Which one of the following does not call for the removal of the name from the State Medical Register?
    A. Death of the doctor
    B. De-recognition of his degree
    C. Fraudulent entry
    D. Professional death sentence

14. Which one of the following is an exception to the doctrine of Professional Secrecy?
    A. Privileged communication
    B. Res judicata
    C. Therapeutic privilege
    D. Therapeutic misadventure

[Answers: 11 - A, 12 - C, 13 - B, 14 - A]

IDENTIFICATION

1. Cephalic index is used to determine the
   A. Height
   B. Cranial capacity
   C. Race
   D. Sex

2. Which one of the pelvic feature determines the sex with accuracy in eight year old girl?
   A. Shape of the obturator foramen
   B. Greater sciatic notch
   C. Pre-auricular sulcus
   D. Sub pubic angle

3. The first permanent tooth to erupt is the
   A. Central incisor
   B. Lateral incisor
   C. First premolar
   D. First molar

4. The second permanent molar generally erupts at the age of
   A. 7 to 8 years
   B. 9 to 10 years
   C. 12 to 14 years
   D. 17 to 25 years

5. At the age of 9 years, the numbers of permanent teeth present are
   A. 12
   B. 16
   C. 20
   D. 24

[**Answers:** 1 – C, 2 – B, 3 – D, 4 – C, 5 – B]

6. A person attains the age of majority on completion of
    A. 16 years
    B. 18 years
    C. 21 years
    D. 24 years

7. The minimum age of marriage for a girl is
    A. 16 years
    B. 18 years
    C. 21 years
    D. 24 years

8. In which one of the following bones, the ossification center appears at the age of viability?
    A. Calcaneum
    B. Talus
    C. Cuboid
    D. Lower end of the femur

9. The foolproof method of identification is
    A. Dactylography
    B. DNA fingerprinting
    C. Anthropometry
    D. Photography

10. The most common pattern of fingerprint is
    A. Arch
    B. Loop
    C. Whorl
    D. Composite

[**Answers:** 6 – B, 7 – B, 8 – B, 9 – A, 10 – B]

11. Estimation of the age of an adult over 21 years by the examination of the teeth is known as
    A. Bertillon's method
    B. Galton's method
    C. Gustafson's method
    D. Locard's method

12. 'Davidson's body', the feminine trait is demonstrated from the
    A. Buccal mucosa
    B. Blood
    C. Cerebrospinal fluid
    D. Urine

13. Which one of the following cranial suture unites first?
    A. Coronal
    B. Frontal
    C. Metopic
    D. Sagittal

14. If the tattoo mark has disappeared, the pigments may still be traced in the
    A. Underlying muscles
    B. Subcutaneous tissues
    C. Regional lymph nodes
    D. All of the above

15. Which one of the following cannot be a deciduous tooth?
    A. Incisor
    B. Canine
    C. Pre molar
    D. Molar

[**Answers:** 11 – C, 12 – B, 13 – C, 14 – C, 15 - C]

16. 'Corpus delicti' means the body
    A. Of the victim
    B. Of evidence
    C. Of the accused
    D. That is ready for autopsy

17. Pre auricular sulcus is prominent in
    A. Females accustomed to sexual intercourse
    B. A female who is pregnant
    C. A female who has borne at least one child
    D. Virgins

18. Pre auricular sulcus is formed due to the pressure exerted by the
    A. Anterior sacro-iliac ligament
    B. Horns of the uterus
    C. Round ligament of uterus
    D. None of the above

19. Which one of the following carpal bone appears between 10 to 12 years?
    A. Lunate
    B. Pisiform
    C. Triquetral
    D. Trapezoid

20. The last temporary tooth to erupt is
    A. Canine
    B. First molar
    C. Second molar
    D. Third molar

[**Answers:** 16 - B, 17 - C, 18 - A, 19 – B, 20 - C]

THANATOLOGY

1. If the dead body is flaccid and cold to touch, which one of the following interpretation is correct?
   A. The body is dead for more than 36 hours
   B. The body is in a state of somatic death.
   C. Organs are viable for transplantation
   D. The muscles are in a stage of primary flaccidity

2. In the areas of contact flattening
   A. Rigor mortis does not occur
   B. Post-mortem lividity does not develop
   C. Putrefaction is delayed
   D. Adipocere formation is faster

3. Which one of the inference drawn from the post-mortem examination finding is incorrect?
   A. If the stomach has easily recognizable undigested rice particles, the time since death is less than 2 hours.
   B. If the rigor mortis is present only in lower limb, the upper part is in a state of secondary flaccidity.
   C. Time since death is less than 8 hours if post-mortem lividity blanches on pressure.
   D. Marbling if found, the postmortem interval is 36 to 48 hours.

4. In the dead body if left elbow joint is easily flexible and the right elbow is stiff, the inference is
   A. Person is left handed therefore the rigor has not yet developed.
   B. That the left limb was paralyzed when he was alive.
   C. That the rigor in the left limb is broken during the handling of the body and therefore it is an artifact.
   D. That the rigor is in a phase of passing.

[**Answers:** 1 - A, 2 - B, 3 - A, 4 - C]

5. The state of suspended animation can be seen in
   A. New born
   B. Hypothermia
   C. Electrocution
   D. All of the above

6. Post-mortem caloricity may be seen in deaths due to
   A. Septicemia
   B. Strychnine poisoning
   C. Tetanus
   D. All of the above

7. The color of the post-mortem lividity is bright red in deaths due to
   A. Carbon monoxide
   B. Cyanide
   C. Phosphorous
   D. Hydrogen sulphide

8. The color of the post-mortem lividity is cherry red in deaths due to
   A. Carbon monoxide
   B. Cyanide
   C. Phosphorous
   D. Hydrogen sulphide

9. Post-mortem lividity helps in all of the following except
   A. Determining the change in the position of the body after fixation
   B. Determination of the cause of death
   C. Estimation of the time since death
   D. Determination of the place of death

10. The chief organism that is responsible for putrefaction is
   A. Clostridium welchii
   B. E. coli
   C. Proteus
   D. Pseudomonas

11. The first sign of putrefaction seen externally is in the
    A. Chest wall
    B. External genitalia
    C. Right iliac fossa
    D. Peri orbital region

12. Casper's dictum
    A. Deals with the principle of exchange of matter when two objects touch each other
    B. Distinguishes ante mortem hanging from postmortem suspension
    C. Helps in estimating the percent of burns
    D. Relates to rate of putrefaction in different media

13. Warm humid atmosphere, still air, shade and abundant fat in the body helps in
    A. Adipocere formation
    B. Brightening the post-mortem lividity
    C. Faster development of rigor mortis
    D. Mummification

14. In which of the following poisoning the putrefactive rate is reduced?
    A. Arsenic
    B. Phenol
    C. Strychnine
    D. All of the above

[**Answers:** 11 - C, 12 - D, 13 - A, 14 - D]

MECHANICAL INJURY

1. Lands and Grooves inside the barrel of a firearm help to
   A. Narrow down the spread of pellets
   B. Impart spin to the pellet
   C. Identify make of the firearm after examining the projectile
   D. Increase the cosmetic appearance of the firearm

2. Entrance wound of a rifled firearm shot from distant range is characterized by
   A. Singeing of hairs around the wound
   B. Tattooing of the adjacent skin
   C. Only inner grease collar and outer abrasion collar
   D. Everted margins

3. Choking in a firearm
   A. Is constriction of the terminal part of the muzzle end
   B. Increases the speed of the bullet
   C. Is constriction of the proximal part of the breech end
   D. Means that the bullets are stuck inside

4. Black powder used as propellant in the firearms contains
   A. Charcoal, potassium nitrate and nitrocellulose
   B. Charcoal, potassium nitrate and nitroglycerine
   C. Charcoal, potassium nitrate and sulphur
   D. Charcoal, nitroglycerine and sulphur

5. The bullet whose nose is sawn off and mushrooms out on hitting the target is known as
   A. Dum Dum bullet
   B. Frangible bullet
   C. Ricochet bullet
   D. Tandem bullet

[**Answers:** 1 – C, 2 – C, 3 – A, 4 – C, 5 - A]

6. A ricochet bullet is the one that
    A. Explodes on hitting the target
    B. Hits an unintended target after hitting the primary target aimed at
    C. Hits an intermediary target, gets deflected and then hits the unintended target
    D. Hits an intermediary target, gets deflected and then hits the primary target aimed at

7. Which one of the following is not a part of the shotgun cartridge?
    A. Bullet
    B. Cardboard wad
    C. Felt wad
    D. Propellant

8. Which one of the following feature is not seen at the entry wound caused by the bullet that has passed through the wearing apparel?
    A. Abrasion collar
    B. Circular shape of the wound
    C. Grease collar
    D. None of the above

9. Bullet recovered from the body is sent to the ballistic expert in a
    A. Paper packet put in a padded box without washing
    B. Plastic cover after thoroughly cleaning and shade drying
    C. Plastic cover put in a padded box without washing
    D. Plastic cover after drying it in the sun

10. The exit wound at the back of the victim who was killed at point blank range shows
    A. Burning of the skin
    B. Blackening
    C. Tattooing
    D. None of the above

[**Answers:** 6 - C, 7 – A, 8 – C, 9 – A, 10 - D]

11. Souvenir bullet is one that
   A. Has no penetrating power
   B. Is embedded in the tissues for long
   C. Is unjacketed
   D. The victim desires to keep for himself after its removal from his body, as a souvenir

12. Dermal nitrate test is used to detect the gunpowder residue in the
   A. Culprits hand
   B. Exit wound
   C. Entry wound
   D. Victim's blood

13. Exit wound is characterized by all of the following except
   A. Everted margins
   B. Protrusion of fat
   C. Tattooing at the edges
   D. Wound bigger than the size of the bullet

14. The lands and grooves inside the barrel of the firearm
   A. Imparts spin to the bullet
   B. Prevents the early spread of the pellets
   C. Is meant to reduce the weight of the firearm
   D. Prevents ricocheting of the bullet

15. In which of the following order the contents of the shotgun cartridge is arranged?
   A. Propellant, cardboard wad, felt wad, cardboard wad, pellets and card board wad
   B. Propellant, felt wad, cardboard wad, pellets, and card board wad
   C. Propellant, cardboard wad, felt wad, pellets and cardboard wad
   D. Primer, cardboard wad, Propellant, cardboard wad, pellets and cardboard wad

[**Answers:** 11 – B, 12 – A, 13 – C, 14 – A, 15 - C]

16. Which one of the following statement regarding the bullet injury to the skull is not correct?
    A. Inner table at the entry is beveled
    B. Outer table at the exit is beveled
    C. Inner table at the exit wound is beveled
    D. Outer table at the entry is clean cut

17. A 'Brush burn' is
    A. An extensive burn
    B. A burn produced by electricity
    C. Graze abrasion
    D. Pressure abrasion

18. Which one of the following is not an example of pressure abrasion?
    A. Ligature mark in hanging
    B. Ligature mark in strangulation
    C. Teeth bite marks
    D. Fingernail marks

19. A blow of moderate violence may produce a comparatively small bruise, if the
    A. Tissues are loose
    B. Tissues are firm and fibrous
    C. Tissues are overlying the bones
    D. Patient is anaemic

20. The age of the bruise can be determined by
    A. Its size
    B. The amount of extravasated blood
    C. The changes in the colour
    D. All of the above

[**Answers:** 16 – C, 17 – C, 18 – D, 19 – A, 20 - C]

21. Which one of the following statement is not correct for an incised wound?
    A. The length is more than the width
    B. The edges are clean cut
    C. Deeper at the beginning
    D. Is spindle shaped

22. Chop wounds are caused by
    A. Blunt-pointed weapon
    B. Light sharp cutting weapon
    C. Heavy sharp cutting weapon
    D. Broken piece of glass

23. Incised wounds on the genitals are usually
    A. Self-inflicted
    B. Fabricated
    C. Homicidal
    D. Accidental

24. The depth of a stab wound
    A. Is equal to the length of the blade
    B. Is less than the length of the blade
    C. Is greater than the length of the blade
    D. Has no relation to the length of the blade

[**Answers:** 21 – D, 22 – C, 23 –A, 24 - D]

REGIONAL INJURY

1. The term open head injury is used to connote
   - A. Blunt force trauma to the brain
   - B. Tear in the scalp
   - C. Fracture of the skull
   - D. Tear in the dura

2. Which one of the following fractures is not a feature of adult skull?
   - A. Depressed fracture
   - B. Fissured fracture
   - C. Pond fracture
   - D. Gutter fracture

3. Fracture a-la signature is caused by
   - A. Blast effect of bomb explosion
   - B. Fall from a height
   - C. Heavy weapon with a small striking surface
   - D. Light weapon with a large striking surface

4. A blunt force trauma on the head resulting in the separation of the sutures is known as
   - A. Comminuted fracture
   - B. Fissure fracture
   - C. Diastatic fracture
   - D. Radiating fracture

5. Fracture not due to direct impact but from transmitted force is
   - A. Fissure fracture
   - B. Gutter fracture
   - C. Pond fracture
   - D. Ring fracture

[**Answers:** 1 - D, 2 - C, 3 - C, 4 - C, 5 – D]

6. Gutter fracture is caused when
    A. A person falls head down in to the gutter
    B. A bullet grazes the outer table of the skull
    C. The head is hit with a corrugated, galvanized heavy iron rod
    D. The primary and the secondary impact injuries are caused by the same vehicle

7. Cerebral concussion has all the following features except
    A. Contusion of the brainstem
    B. Retrograde amnesia
    C. Functional paralysis of the muscles
    D. Transient loss of consciousness

8. The term 'Organization of blood clot' is associated with
    A. Epidural haemorrhage
    B. Subdural haemorrhage
    C. Intra cerebral haemorrhage
    D. Subarachnoid haemorrhage

9. Rupture of the berry aneurysm leads to
    A. Extradural haemorrhage
    B. Sub arachnoid haemorrhage
    C. Intracerebral haemorrhage
    D. Subdural haemorrhage

10. In a victim of road traffic accident, bleeding from the ear indicates
    A. Bleeding in to the subarachnoid space
    B. Extensive contre coup injury to the brain
    C. Fracture of the middle cranial fossa
    D. Tear in the middle meningeal artery

[**Answers:** 6 - B, 7 - A, 8 - B, 9 – B, 10 - C]

11. Railway spine is the spinal cord
    A. Concussion
    B. Crush injury
    C. Contusion
    D. Laceration

12. Punch drunk is
    A. Consumption of alcohol mixed with knock out drops
    B. Driving under the influence of alcohol
    C. Repeated head injury sustained by the boxers affecting the extra pyramidal system
    D. Sustaining head injury under the influence of alcohol

[**Answers:** 11 – A, 12 - C]

TRANSPORTATION INJURY

1. In a road traffic accident, secondary impact injuries are sustained by the
   A. Front seat occupants of the motor car
   B. Pillion rider of the motor bike
   C. Pedestrians
   D. Passengers of a moving train

2. Flail chest, one of the outcome of vehicular accident is seen in
   A. Back seat occupant
   B. Front seat occupant
   C. Pedestrian
   D. The driver

3. In a vehicular accident if the distance between the heel and the fracture site of the leg bone of the pedestrian is less than the Ground - Bumper distance of the apprehended vehicle
   A. The apprehended vehicle is not the one involved in the accident
   B. It indicates driver's full effort to stop the vehicle by applying the brakes
   C. The fracture is not due to vehicular impact but due to fall on the ground
   D. It indicates that the pedestrian jumped upwards to   avoid the vehicle hitting him

4. All the statements in relation to flail chest given below are true except that there is
   A. Expansion of the chest during expiration
   B. Fractures of the ribs on either side of the sternum
   C. Ineffective movements of the chest wall
   D. Severance of phrenic nerve leading to diaphragmatic paralysis

[**Answers:** 1 – C, 2 – D, 3 – B, 4 – D, 5 - D]

5. The lesion caused by sudden hyper extension and hyper flexion of the head is known as
   A. Concussion of the brain
   B. Railway spine
   C. Fracture of the hyoid bone
   D. Whiplash injury

6. Abraded contusion on the left parietal area, grazed abrasion on the left upper arm in a victim of road traffic accident are grouped under
   A. Primary impact injuries
   B. Secondary injuries
   C. Secondary impact injuries
   D. Simple injuries

7. Whiplash injury is mostly associated with the
   A. Cyclist
   B. Pillion rider
   C. Pedestrian
   D. Occupant of a vehicle

[**Answers:** 5 – D, 6 – B, 7 - D]

THERMAL INJURY

1. When the entire back of the trunk of a person is burnt, the body surface area involved is
    - A. 9%
    - B. 18%
    - C. 27%
    - D. 36%

2. Which one of the following is not an artifact associated with death due to burns?
    - A. Cherry red blood
    - B. Heat fracture
    - C. Heat rupture
    - D. Heat hematoma

3. Arborescent markings when present indicates the death as due to
    - A. Fall in boiling water
    - B. Gas stove burst
    - C. Heat stroke
    - D. Lightning

4. The commonest cause of death from electrical injury is due to
    - A. Aspiration of the stomach contents due to violent spasms
    - B. Ventricular fibrillation
    - C. Respiratory muscle paralysis
    - D. Vagal inhibition

5. Newborns are more vulnerable to ill effects of cold because
    - A. Heat regulation center is not well developed
    - B. The vasomotor reflexes are still not well established
    - C. The surface area is greater in relation to the body
    - D. All of the above

[**Answers:** 1 – B, 2 – A, 3 – D, 4 – B, 5 - D]

6. Which one of the following regarding the heat rupture is not true?
    A. Bleeds profusely
    B. Has irregular margins
    C. Margins shows no bruising
    D. Shows intact nerves crossing the depth of the tissues

7. The pathognomonic feature of death due to ante mortem burns is
    A. Blisters on the skin
    B. Singeing of the hairs
    C. Pugilistic posture of the body
    D. Soot particles in the trachea

8. Which one of the following is not an indicator of death from antemortem burns?
    A. Cherry red colour of the blood
    B. Pugilistic posture of the body
    C. Presence of soot particles in the trachea
    D. Pus found in the wounds

9. Palisade arrangement of the cells with elongated nuclei is a feature of
    A. Acid burn
    B. Flame burn
    C. Electric burn
    D. Rope burn

10. Hypothermia is diagnosed when the core body temperature falls below
    A. 30 C
    B. 35 C
    C. 37 C
    D. 38 C

[**Answers:** 6 – A, 7 – D, 8 – B, 9 – C, 10 - B]

11. 'Pugilistic attitude' is associated with all of the following except
    A. It indicates exposure to great amount of heat
    B. It is due to coagulation of muscle protein
    C. It occurs only in ante-mortem burns
    D. Flexors are more affected than extensors

12. Curling's ulcer in the victims of burns is usually seen in
    A. Oesophagus
    B. Stomach
    C. Duodenum
    D. Colon

13. Which one of the following is not common to both burn and scald injury?
    A. Blisters
    B. Erythema
    C. Necrosis of tissues
    D. Singeing of hairs

[**Answers:** 11 – C, 12 – C, 13 - D]

MEDICO-LEGAL ASPECTS OF WOUND

1. A stab injury in to the thorax is grievous hurt because it
   A. Emasculates the person
   B. Endangers the life of the person
   C. Heals with scar formation thus causing   permanent disfiguration
   D. Requires 20 days of hospitalization

2. What should be the minimum percentage of body surface involved to constitute grievous hurt?
   A. One third of the body surface
   B. Two thirds of the body surface
   C. Three fourths of the body surface
   D. More than 50% of the body surface

3. Legally speaking an offer of threat with the ability to do so constitute
   A. Assault
   B. Battery
   C. Hurt
   D. Injury

4. As per section 320 IPC, 'member' does not include
   A. Finger
   B. Hand
   C. Hair
   D. Penis

5. The final decision to consider an injury as grievous lies on
   A. Doctor
   B. Investigating officer
   C. Judge
   D. Public prosecutor

[**Answers:** 1 – B, 2 – A, 3 – A, 4 – C, 5 - C]

6. Killing a person under gross provocation is an example of
    A. Excusable homicide
    B. Homicide not amounting to murder
    C. Homicide amounting to murder
    D. Justifiable homicide

7. The surest sign of an ante-mortem wound is
    A. Bleeding
    B. Coagulation of blood
    C. Positive histo-chemistry
    D. Pus formation

8. Sudden death may ensue from all of the following except
    A. Crush syndrome
    B. Haemorrhage
    C. Injury to the vital organ
    D. Shock

[**Answers:** 6 – B, 7 – D, 8 – A]

MECHANICAL ASPHYXIA DEATH

1. Injury to the spinal cord is always seen in
   A. Accidental hanging
   B. Judicial hanging
   C. Homicidal hanging
   D. Suicidal hanging

2. Which one of the following is not an example of death due to suffocation?
   A. Choking
   B. Gagging
   C. Mugging
   D. Smothering

3. Apoplexy is the term used for cerebral
   A. Anoxia
   B. Concussion
   C. Congestion
   D. Ischemia

4. When any part of the body touches the ground in hanging it is called
   A. Atypical hanging
   B. Complete hanging
   C. Partial hanging
   D. Typical hanging

5. The condition where the drowning fluid penetrates the alveolar walls to enter the tissues and the blood vessels is known as
   A. Hydrostatic lung
   B. Emphysema aquosum
   C. Oedema aquosum
   D. Paultof's haemorrhage

[**Answers:** 1 – B, 2 – C, 3 – C, 4 – C, 5 - B]

6. Hydrocution refers to death due to
    A. Electrocution in water
    B. Vagal inhibition on entry into cold water
    C. Laryngeal spasm due to submersion
    D. Lungs filled with water preventing entry of air.

7. Atypical incomplete hanging means the knot is at the
    A. Back and the feet are touching the ground.
    B. Back and the feet are not touching the ground
    C. Sides and feet are not touching the ground.
    D. Sides and feet are touching the ground.

8. Which one of the following conditions belongs to death from traumatic asphyxia?
    A. Trauma to the cervical spine in judicial hanging.
    B. A bolus of food impaction in the larynx
    C. Death due to stampede
    D. Trauma to the head of a motorcycle rider.

9. Gettler's test is done to distinguish
    A. Ante-mortem drowning from post-mortem drowning
    B. Ante-mortem hanging from postmortem suspension
    C. Freshwater drowning from salt water drowning
    D. Ante-mortem burns from post-mortem burns.

10. Le Facie sympathique
    A. Indicate the ante-mortem nature of hanging
    B. Is facial palsy due to trauma on the temple
    C. Is ptosis seen in neurotoxic snake bites
    D. Is the description of the face of the victim of vitriolage

[**Answers:** 6 – B, 7 – D, 8 – C, 9 – C, 10 - A]

11. Hydrostatic lung
    A. Is passage of water due to hydrostatic pressure into the lungs of a dead body in water.
    B. Is the one in which hydrostatic test does not conclusively prove the live birth
    C. Refers to fulminate pulmonary oedema.
    D. Is one where the lungs are compressed due to excessive fluid in the pleural cavity.

12. Death from 'Café Coronary' is due to
    A. Choking
    B. Coronary insufficiency
    C. Cerebral stroke
    D. Food poisoning

13. Burking is a combination of
    A. Choking and mugging
    B. Mugging and gagging
    C. Gagging and choking
    D. Homicidal smothering and traumatic asphyxia

14. Which one of the following does not come under the category of death due to strangulation?
    A. Bansdola
    B. Gagging
    C. Garroting
    D. Throttling

15. Classical signs of death from drowning, like copious amount of white froth is seen in
    A. Dry drowning
    B. Immersion syndrome
    C. Near drowning
    D. Wet drowning

[**Answers:** 11 – A, 12 – A, 13 – D, 14 – B, 15 - D]

16. The proof of death from ante-mortem drowning lies in the demonstration of diatoms from
    A. Bone marrow
    B. Lungs
    C. Middle ear
    D. Stomach

17. A tension of 2 to 3 kg in the ligature around the neck is required to block the
    A. Jugular veins
    B. Carotid arteries
    C. Trachea
    D. Oesophagus

18. All of the following are the cardinal signs of asphyxia except
    A. Cyanosis
    B. Petechial haemorrhages
    C. Increased capillary permeability
    D. Ventricular fibrillation

19. Lynching is
    A. A form of homicidal hanging
    B. Accidental hanging from the rungs of the ladder
    C. Atypical hanging with multiple ligatures around the neck
    D. Post-mortem suspension of the body

20. Death due to cyanide poisoning is an example of
    A. Anaemic anoxia
    B. Anoxic anoxia
    C. Histotoxic anoxia
    D. Stagnant anoxia

[**Answers:** 16 – A, 17 – A, 18 – D, 19 – A, 20 - C]

21. Café coronary victims are
   A. Children below 16 years
   B. Intoxicated persons
   C. Elderly people above 60 years
   D. Mentally unsound persons

22. Diatoms are
   A. Crustaceans
   B. Non motile bacteria
   C. Unicellular algae
   D. Water dependent fungi

23. Hypoxia leading to the damage to the cement substance between the endothelial cells of the capillaries results in
   A. Atrophy of the organ
   B. Transudation of the fluid
   C. Rupture of the vessel
   D. None of the above

24. The surest sign of death form asphyxia is
   A. Cerebral oedema.
   B. Petechial haemorrhages
   C. Intense cyanosis
   D. None of the above

25. Which one of the following is not related to death from drowning?
   A. Diatom test
   B. Gettler's test
   C. Hydrostatic test
   D. Slide test

[**Answers:** 21 – B, 22 – C, 23 – B, 24 – D, 25 - C]

26. In a case of death due to manual strangulation, dissection of the neck should follow a carefully laid out method to
    A. Avoid the neck structures to drain thoroughly of blood
    B. Differentiate bruises from livid areas
    C. Avoid dissection artifacts
    D. Rule out death due to vagal inhibition

27. Which one of the following finding, if present, is an indicator of death due to ante-mortem hanging?
    A. Nail marks on either side of the neck
    B. Dribbling of the saliva from the angle of the   mouth
    C. Postmortem Lividity on the lower part of the legs and forearms
    D. Noose above the level of the thyroid and the knot exactly at the back of the head

28. Fixation of the chest preventing the respiratory movement leading to death is called
    A. Auto erotic sexual asphyxia
    B. Mugging
    C. Burking
    D. Traumatic asphyxia

29. Post-mortem lividity on the chest, entire upper and lower limbs on the body found hanging suggest
    A. Ante-mortem hanging
    B. Postmortem suspension
    C. Atypical hanging
    D. Struggle before hanging

30. Death due to accidental entry of a piece of meat while dining is
    A. Cafe coronary
    B. Choking
    C. Gagging
    D. Restaurant asphyxia

[Answers: 26 – C, 27 – B, 28 – D, 29 – B, 30 - B]

IMPOTENCY AND STERILITY

1. The commonest cause of impotence in males is
   A. Absent genitalia
   B. Diabetes mellitus
   C. Hypo-pituitarism
   D. Psychogenic

2. Which one of the following is not an indication for 'Artificial Insemination'?
   A. Genetic defect in the husband
   B. Rh incompatibility
   C. Sterile husband
   D. Vaginismus

3. Consent for permanent sterilization in a woman must be obtained from
   A. Husband
   B. Wife
   C. Both husband and wife
   D. Legal guardian

4. All the following constitute legal problems for artificial insemination in India except
   A. Adultery
   B. Nullity of marriage
   C. Legitimacy
   D. Inheritance of property

5. Oligospermia could occur in all of the following conditions except
   A. Cryptorchidism
   B. Drug abuse
   C. General debilitating diseases
   D. Prolonged abstinence

[**Answers:** 1 – D, 2 – D, 3 – B, 4 – A, 5 - D]

6. Asthenospermatogenia means
    A. Failure to produce living sperms
    B. Failure to produce adequate number of sperms
    C. Defect in sperm maturation
    D. Defect in sperm movements

[**Answers:** 6 – D]

VIRGINITY, PREGNANCY AND DELIVERY

1. The hymen often escapes injury in a case of sexual assault on a child because it is
   A. Deeply seated
   B. Easily distensible
   C. Tough in nature
   D. Undeveloped

2. All of the following are likely to rupture the hymen except
   A. Horse riding
   B. Per vaginal examination
   C. Sexual intercourse
   D. Use of sanitary tampons

3. Earliest diagnosis of pregnancy can be made by
   A. Echocardiography
   B. Foetoscope
   C. Radiography
   D. Ultrasonography

4. An 'Affiliation case' refers to
   A. Adultery
   B. Non consummation of marriage
   C. Paternity dispute
   D. Sexual Paraphylia

5. Which one of the following is not a conclusive sign of pregnancy?
   A. Foetoscopy
   B. Immunological tests
   C. Ultrasonography
   D. Radiography

[Answers: 1 – A, 2 – A, 3 – D, 4 – C, 5 - B]

6. A supposititious child is one who
    A. Is illegitimate
    B. Is born after the death of the father
    C. Does not belong to the woman claiming to be its mother
    D. Has been abandoned by the parents

7. Ova of the same ovulatory period getting fertilized by separate acts of coitus is referred to as
    A. Superfoetation
    B. Superfecundation
    C. Pseudocyesis
    D. Foetus papyraceous

8. The vaginal discharge in the first few days after the delivery is referred to as
    A. Lochia rubra
    B. Lochia serosa
    C. Lochia alba
    D. Lochia nigra

9. Evidence of torn hymen
    A. Is conclusive proof of rape
    B. Is proof of sexual 9intercourse
    C. Is indicative of masturbation
    D. None of the above

10. The word 'consummation' in marriage means
    A. Buccal coitus
    B. Successful erection on the first night
    C. Sexual intercourse
    D. Wife becoming pregnant

[**Answers:** 6 – C, 7 – B, 8 – A, 9 – D, 10 - C]

11. The definitive sign of pregnancy is
    A. Hearing the heart sounds of the foetus
    B. Increase in the size of the abdomen
    C. Positive HCG test
    D. Quickening

12. Pseudosyesis is found in
    A. Childless wife nearing menopause
    B. In males who are impotent to a particular woman
    C. Unmarried woman in whom contraceptives has failed
    D. Young married woman with illicit sexual relation with a colleague

13. Cribriform hymen refers to
    A. Cresentic opening in the hymen
    B. Sieves in the hymen
    C. Septum in the hymen
    D. None of the above.

14. Which one of the following is not a feature seen on the baby born of precipitate labour?
    A. Intracranial haemorrhages
    B. Pond fracture
    C. Prominent caput succedaneum
    D. Torn umbilical cord

[**Answers:** 11 – A, 12 – A, 13 – B, 14 – C]

ABORTION

1. Which one of the following is not a ground for termination of pregnancy as per MTP Act?
   A. Child likely to be deformed
   B. Failure of contraception in a married woman
   C. Pregnancy resulting from rape
   D. Cephalopelvic disproportion

2. Termination of pregnancy under 'Eugenic ground' is permitted only if the length of pregnancy has not exceeded
   A. 16 weeks
   B. 20 weeks
   C. 24 weeks
   D. 28 weeks

3. The maximum duration of pregnancy up to which abortion can be done under 'Therapeutic ground' is
   A. 12 weeks
   B. 20 weeks
   C. 28 weeks
   D. No time limit

4. Opinion of the second doctor is mandatory to terminate pregnancy if the duration is
   A. Less than 12 weeks
   B. More than 12 weeks
   C. Between 12 to 28 weeks
   D. More than 28 weeks

[**Answers:** 1 – D, 2 – B, 3 – D, 4 - B]

5. When the criminal abortion is performed with the consent of the mother, the liability for punishment is for
   A. The mother only
   B. The doctor only
   C. Both the mother and the doctor
   D. None of them

6. Sudden death form vagal inhibition in criminal abortion is due to
   A. Dilatation of the cervix
   B. Perforation of the vault of the uterus
   C. Perforation of the vaginal wall
   D. Perforation of the hymen

7. Legally the term 'abortion' refers to the expulsion of the product of conception
   A. At any time before full term
   B. Before 1st trimester
   C. Before 2nd trimester
   D. Before 3rd trimester

8. According to the MTP Act termination of pregnancy is permitted
   A. In whomever the pregnancy is a result of contraceptive failure
   B. Only when the husband consents
   C. Only when a female doctor performs it
   D. When the foetus has congenital abnormalities

[**Answers:** 5 – C, 6 – A, 7 – A, 8 – D]

INFANTICIDE

1. A child born after 28 weeks of gestation and not showing any signs of life after being issued forth is considered to be
   A. Dead born
   B. Still born
   C. Dead in utero
   D. None of the above

2. In a dead-born child, which one of the following is seen on external examination?
   A. Caput succedaneum
   B. Cephalhaematoma
   C. Maceration
   D. Spaulding's sign

3. Maceration is due to
   A. Aseptic autolysis
   B. Decomposition
   C. Putrefaction
   D. Sepsis

4. The organ studied in 'Breslau's second life test' is
   A. Heart
   B. Lung
   C. Stomach
   D. Spleen

[**Answers:** 1 – B, 2 – C, 3 – A, 4 - C]

5. Which one of the following can be ruled out when caput succedaneum is present?
   A. Obstructed labour
   B. Prolonged labour
   C. Precipitate labour
   D. Vertex presentation

6. The role of the doctor in a suspected case of 'Battered baby syndrome' is to
   A. Inform the police
   B. Inform the human right authorities
   C. Persuade the parents not to ill-treat the baby
   D. Warn the parents of the legal consequences

7. The most reliable parameter to assess the maturity of the foetus is the
   A. Growth of the nails
   B. Length of the foetus
   C. Growth of scalp the hair
   D. Ossification centers

8. Which one of the following statements regarding a 7 month old male baby is incorrect?
   A. Has attained the age of viability
   B. Testes will be at the external inguinal ring
   C. Ossification center for the Talus would have appeared
   D. The crown – heel length would be 35 cm

9. Which one of the following change is seen by 24 hours after birth?
   A. Cicatrisation
   B. Obliteration of the umbilical vessels
   C. Mummification of the cord
   D. Red zone of inflammation

[**Answers:** 5 – C, 6 – A, 7 – D, 8 – B, 9 – B]

10. Following are the acts of omission in Infanticide except
    A. Not clearing the air passages after delivery, leading to suffocation
    B. Not protecting the child from exposure to heat and cold
    C. Not seeking assistance during labour, leading to head injury
    D. Strangulating the child with the umbilical cord

[**Answers:** 10 – D]

SEXUAL OFFENCE

1. The minimum age of consent for valid sexual intercourse by a male is
   A. 12 years
   B. 14 years
   C. 16 years
   D. None of the above

2. The conclusive proof of rape on a virgin is
   A. Presence of semen in the vagina
   B. Presence of tears in the hymen with fresh bleeding
   C. Presence of smegma bacilli, semen and the hymenal tears with fresh bleeding
   D. None of the above

3. In which one of the following, consent for sodomy is legally valid?
   A. When the consenting girl is over 16 years of age
   B. When she is his own wife
   C. When the consenting person is a Hijra
   D. None of the above

4. Which one of the following is not an example of sexual paraphilia?
   A. Fetishism
   B. Masochism
   C. Lesbianism
   D. Sadism

5. Consented sexual intercourse on a girl of 15 years is not statutory rape in
   A. Bihar
   B. Karnataka
   C. Kerala
   D. Manipur

[**Answers:** 1 – D, 2 – D, 3 – D, 4 – C, 5 - D]

6. Unnatural sexual offences are punishable under section
   A. 376 IPC
   B. 377 IPC
   C. 497 IPC
   D. 497 Cr. PC

7. In the absence of sperms, the semen is detected by
   A. Demonstrating the presence of Acid phosphatase
   B. Lugol's iodine test
   C. Detecting fructose in the vaginal wash
   D. Sweetish but disagreeable odour

[**Answers:** 6 – B, 7 – A]

FORENSIC PSYCHIATRY

1. Which one of the following sections of IPC defines the criminal responsibility of an insane?
   A. Section 82
   B. Section 84
   C. Section 85
   D. Section 87

2. A false but a firm belief in something which is not a fact is called
   A. Delusion
   B. Delirium
   C. Illusion
   D. Hallucination

3. An imaginary suffering of a serious illness making a person to go from hospital to hospital for unnecessary investigation and treatment is
   A. Anxiety neurosis
   B. Depressive psychosis
   C. Munchausen's syndrome
   D. Korsakoff's psychosis

4. A false perception without an external stimulus is known as
   A. Hallucination
   B. Delirium
   C. Delusion
   D. Illusion

5. Kleptomania is
   A. An impulsive disorder
   B. An obsessive compulsive disorder
   C. Personality disorder
   D. Somatoform disorder

[**Answers:** 1- B, 2 – A, 3 – C, 4 – A, 5 - A]

6. The normal range of 'Intelligent Quotient' is
    A. 50 to 70
    B. 70 to 90
    C. 90 to 110
    D. 110 to 130

7. Delirium tremens is generally seen in
    A. Alzheimer's disease
    B. Chronic alcoholism
    C. Cocaine intoxication
    D. Schizophrenia

8. Tactile hallucinations are seen in
    A. Datura poisoning
    B. Cannabis poisoning
    C. Morphine
    D. LSD

9. Mc Naughten of Mc Naughten's rule was
    A. An insane criminal
    B. An eminent judge
    C. A psychiatrist
    D. An eminent defense lawyer

10. Which one of the following is not related to criminal responsibility?
    A. Curren's rule
    B. Durrham's rule
    C. Hasse's rule
    D. Mc Naughten's rule

11. A psychopath is a person who is
    A. Always aggressive and violent
    B. Having a personality disorder
    C. Mentally retarded
    D. Suffering from psychosis

[**Answers:** 6 – C, 7 – B, 8 – A, 9 – A, 10 – C, 11 - B]

12. Which one of the following does not induce hallucination?
    A. Cocaine
    B. Cannabis
    C. Morphine
    D. Phenobarbitone

13. All of the following are essential requirements of 'Testamentary Capacity' except
    A. Not subjected to undue influence
    B. Not being blind and deaf
    C. Minimum age 18 years
    D. Sound mind

14. A rope being interpreted as snake is an example of
    A. Delusion
    B. Hallucination
    C. Illusion
    D. Obsession

15. Insertion of a thought overriding the internal resistance is known as
    A. Automatism
    B. Impulse
    C. Obsession
    D. Shamming

16. The ability to write a valid will is called
    A. Documentary ability
    B. Evidentiary competence
    C. Testamentary capacity
    D. Testimonial ability

[**Answers:** 12 – D, 13 – B, 14 – C, 15 – C, 16 - C]

17. Under the Mental Health Act, the insane person is referred to as
    A. Lunatic
    B. Mentally ill person
    C. Mentally disturbed person
    D. Mentally abnormal person

18. An irresistible desire to alcohol in excess is known as
    A. Dipsomania
    B. Alcoholism
    C. Pathological intoxication
    D. Punch drunk syndrome

19. Mental unsoundness at the time of marriage is a ground for
    A. Divorce
    B. Nullity of marriage
    C. Compensation for damages
    D. Alimony

20. Which one of the following is not an affective disorder?
    A. Bipolar disorder
    B. Anxiety neurosis
    C. Depressive disorder
    D. Maniac disorder

21. Somnolentia refers to
    A. Acute insomnia
    B. A condition midway between sleep and wakefulness
    C. Sleep walking
    D. Hypnotic trance

[**Answers:** 17 – B, 18 – A, 19 – B, 20 – B, 21 - B]

22. A state of akinesis and mutism with relative preservation of conscious awareness is called
    A. Stupor
    B. Delirium
    C. Catharsis
    D. Catatonia

[**Answers:** 22 – A]

BLOOD STAINS

1. The micro chemical test for blood showing crystals of Haemochromogen under the microscope is known as
    A. Benzidine test
    B. Kastle-Meyer test
    C. Takayama test
    D. Teichmann's test

2. Precipitin test identifies the
    A. Blood stain
    B. Seminal stain
    C. Species of origin
    D. Vaginal epithelium

3. Davidson bodies, a feminine trait seen in the blood is found in the
    A. Basophils
    B. Eosinophils
    C. Monocytes
    D. Neutrophils

4. Chances of excluding the putative father by blood groups is highest in
    A. HLA system
    B. Red cell antigens
    C. Red cell enzyme polymorphisms
    D. Serum protein polymorphisms

5. In persons who are 'secretors' the agglutinogens of the ABO system are present in all the body fluids except
    A. Cerebrospinal fluid
    B. Saliva
    C. Tears
    D. Vitreous humor

[Answers: 1 – C, 2 – C, 3 – D, 4 – A, 5 - A]

6. The ABO system of blood group is based on the
    A. Red cell antigens
    B. Red cell enzyme
    C. Serum proteins
    D. White cell antigens

[**Answers:** 6 – A]

www.ingramcontent.com/pod-product-compliance
Lightning Source LLC
Chambersburg PA
CBHW021922170526
45157CB00005B/2141